CARD CAPERS

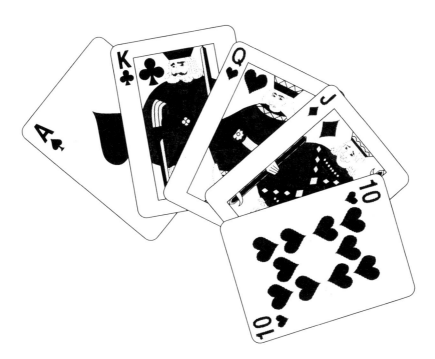

Developing Mathematics from Playing Cards

by Paul Swan

Card Capers first published 1998, reprinted, 2001, 2002, 2003, 2005, 2007

Author: Paul Swan

Copyright © A-Z Type

ISBN 0 9585632 0 9

Printed by Success Print

The author may be contacted at: pswan@iprimus.com.au

Some other books by the same author

I sometimes self publish and at other times I write for publishers. Here is a listing of my current titles.

Published by A–Z Type

  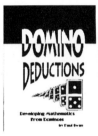 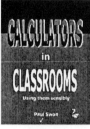  

| Dice Dilemmas | Dice Dazzlers | Domino Deductions | Calculators | Tackling Tables | Turn & Learn |

Other books by the same author published by RIC Publications (www.ricgroup.com.au)

   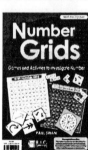  

| Motivational Maths | Maths Investigations | Check Your Work | Number Grids | Maths Terms & Tables | Mastergrids |

| Patterns Middle | Patterns Upper | Unifix | Pattern Blocks | Base Ten Blocks | Maths Facts, Fun, Tricks & Trivia |

Contents

Teachers' Guide 4

Symmetrical Cards 5

Spotted Cards 6

Cut and Predict 7

Magic Cards 8

Boxed Cards 9

Classifying Cards 10

Card Layout 13

Card Conundrum 14

The Cards of Hanoi 15

Shuffle the Cards 16

Card Puzzle 17

Deck Detective 18

Card Games 19

Beat That 20

Card Bingo 21

Up and Down 22

Fish 23

Snap +/− 1 24

Counting Cards 25

Make 10 26

Make 10 Again 27

Flipper 28

I Spy 30

Thirty One 31

Place Value Pack 32

Multiple Madness 33

Practice Pack 34

Getting Closer 35

99 or Bust 36

Hit the Deck 37

On Target 38

Fast Facts 39

HiLo 40

Make my Number 41

Card Count 42

Card Nim 43

Secret Pairs 44

Calculate a Digit 46

Card Calculations 47

Answers 48

Teachers' Guide

Playing Cards are:

♠ familiar to most children

♠ non-threatening

♠ highly motivational and

♠ found in most homes

which makes them an ideal piece of material to use as the basis for work at school and at home.

Playing cards are well suited to developing

♠ number work

♠ problem solving

♠ classifying

♠ probability and

♠ symmetry

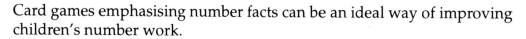

Card games emphasising number facts can be an ideal way of improving children's number work.

To simplify organisation, the book has been divided into two sections; problems and puzzles and card games. The Year suitability and content of each activity is given by the code on the top right of each page.

J = junior primary,

M = middle primary,

U = upper primary,

S = space,

N = number,

PS = problem solving,

Pr = probability,

PV = place value and

MC = mental computation. This category has been broken down even further into +, – and x.

The materials and suggested organisational structure for each game are given to help teachers prepare for playing the games outlined in the book. It has been assumed that the children have an idea of how to shuffle and deal cards as well as other typical card conventions such as dealing to the left in a clockwise rotation and dealing to oneself last.

Symmetrical Cards

♦ Look at the symbols on the numbered cards in a standard deck (i.e. ♣ ♦ ♥ ♠).

♦ Notice how they are arranged in a symmetrical pattern.

♦ Which cards have line symmetry?

♦ Which cards have rotational symmetry?

♦ Which cards have both rotational and line symmetry?

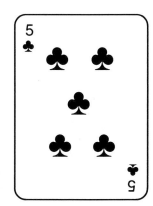

♦ Design some playing cards to represent 11, 12 and 13.

11

12

13

Spotted Cards

♠   How many spots (i.e. ♣ ♦ ♥ ♠ symbols) are there on a deck of cards?

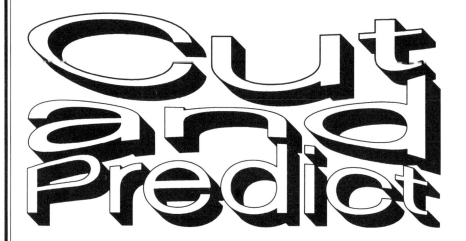

- Remove the Jokers and shuffle a pack of playing cards. Cut the deck and record the colour of the card.

- Try this twenty times.

- Write about what you notice.

- If you were to try this fifty, one hundred or two hundred times, how many red cards would you expect to turn up?

- Repeat the experiment, but this time, note which suit appears.

- Predict the number of hearts that you would expect to get from forty, sixty or one hundred cuts from the deck.

Variations

- Consider cutting for picture cards.

♦ Separate an Ace (Ace = one), 2, 3, 4, 5, 6, 7, 8 and 9 from a deck of cards.

♦ Place the cards in a 3 x 3 grid on the table.

e.g.

♦ Find the total of the first row, 7 + 4 + 3 = 14

Find the total of the second row, 2 + 6 + A = 9

Find the total of the third row, 8 + 5 + 9 = 22

♦ Add these three totals.

♦ Collect the nine cards, shuffle them and place them on the table in a different arrangement. Add the totals.

♦ Repeat once more.

What do you notice? Why do you think this happens?

M:MC +

BOXED CARDS

♣ Remove ten cards (Ace to 10) from a deck of cards and form a box.

$(1 + 9 + 8)$

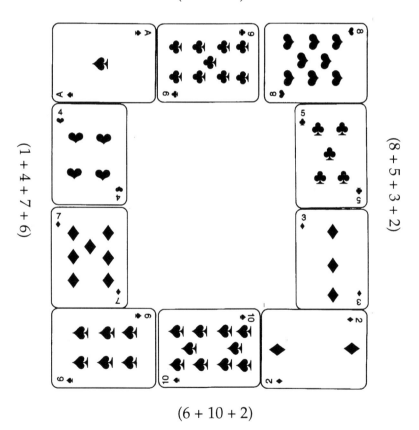

$(1 + 4 + 7 + 6)$

$(8 + 5 + 3 + 2)$

$(6 + 10 + 2)$

♣ Add up the cards on each side of the box.

♣ Try to find another arrangement that produces a sum of eighteen on each side of the box. When adding the values on each side of the box count three cards on the top and bottom and four cards down each side.

♣ Cards may be classified or sorted into various groups. For example children in junior grades can sort and classify cards according to suit, colour, or whether the cards are even or odd. Young children might also place the cards in order according to the number of symbols shown on the card. Later they can use the numerals.

♣ Carroll diagrams (named after Lewis Carroll, the author of Alice in Wonderland) or Venn diagrams may be used to help organise the sorting process.

For example cards may be sorted according to the following criteria.

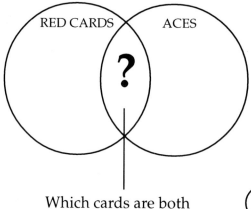

Which cards are both red and aces?

Ace of hearts & Ace of diamonds

♣ *How many cards end up in each location?*

♣ Try sorting the cards according to the following criteria.

	Odd Number	Even Number	Picture Card
R E D			
B L A C K			

Classify the cards according to the following criteria.

1.

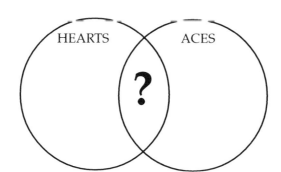

HEARTS   ACES

?

2.   PICTURE   DIAMONDS
            CARDS

?

3.   ODD CARDS   CLUBS

?

4.   PICTURE   BLACK
            CARDS   CARDS

?

5.   CLUBS   PICTURE
                           CARDS

?

6.   TENS   DIAMONDS

?

7.   FOURS   BLACK
                           CARDS

?

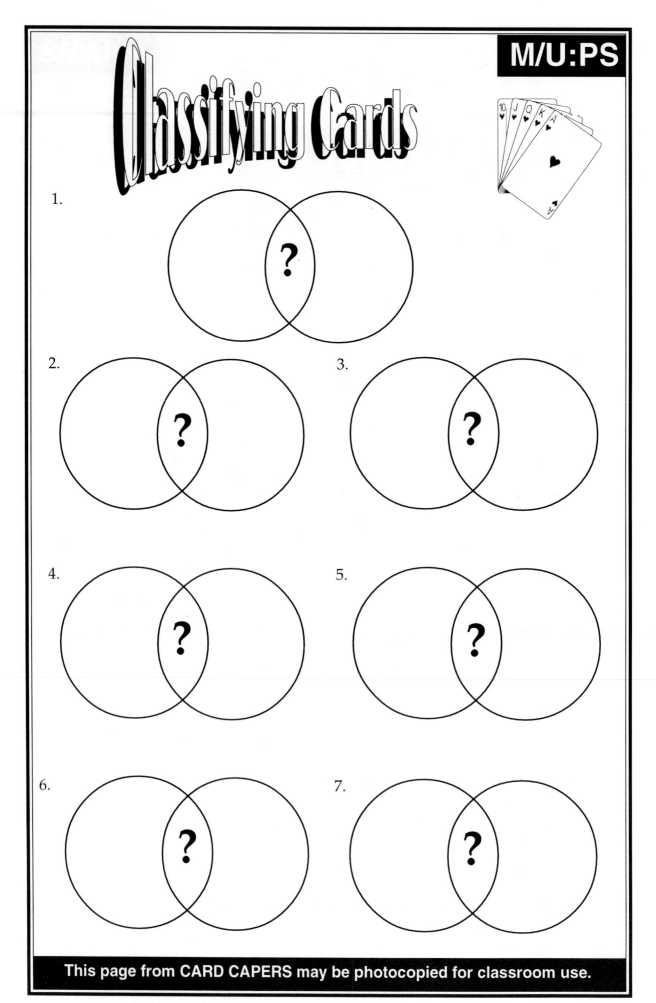

Classifying Cards

1.

2.

3.

4.

5.

6.

7.

♦ Arrange four numbered cards in a 2 x 2 array.

♦ Find the total for each row and column.

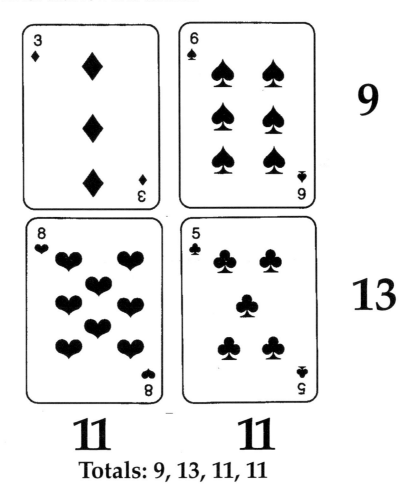

Totals: 9, 13, 11, 11

♦ Try to rearrange the cards so that differents totals are produced.

♦ How many different totals are possible?

♦ How will you know when you have found them all?

♦ Try starting with a 2 x 3 array.

Card Conundrum

◆ Remove all the Aces and picture cards from a deck of cards and form a 4 x 4 array in which each row and each column contains one Ace, King, Queen and Jack.

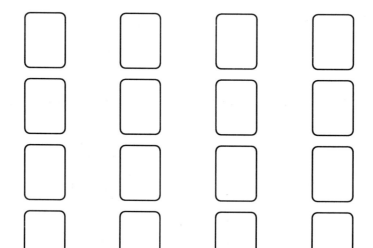

◆ Try again, but this time form a 4 x 4 array in which each row, column and *diagonal* contains one Ace, King, Queen and Jack.

◆ Form a 4 x 4 array in which no two cards of the same value or the same suit appear in the one row, column or diagonal.

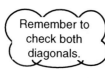

Remember to check both diagonals.

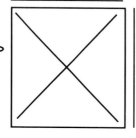

This is really hard. My brain hurts!

The Cards of Hanoi

In the famous "Tower of Hanoi" puzzle, discs are transferred from one peg to another according to the rules that:

♦ only one disc may be moved at a time

♦ a larger disc cannot be placed on top of a smaller disc.

A similar puzzle may be produced using playing cards.

♦ Separate an ace, a 2, a 3 and a 4 from a deck of cards.

♦ The object of the puzzle is to move the four cards from point A to point C.

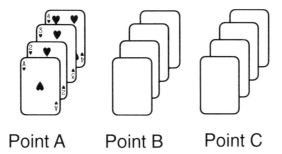

Point A   Point B   Point C

Use the following rules:

♦ A larger valued card cannot be placed on top of a lesser valued card, e.g. a three cannot be placed on top of a two or an ace.

♦ Only one card may be moved at a time. Note cards may be moved directly from point A to point C.

♦ If you find the puzzle hard then remove the four. If you find it easy to solve then try adding a five and a six.

> Caution: With five or six cards this puzzle becomes very difficult and takes a long time to complete.

Shuffle the Cards

♥ Remove nine cards (Ace to 9) from a deck.

♥ Shuffle the nine cards and lay them out, face up, in a 3 x 3 array.

♥ Try to move the cards so they are in descending order from nine to one, as shown below.

♥ You may only move a single card at a time. A card may only be moved on top of another one if it is of a higher value.
E.g.

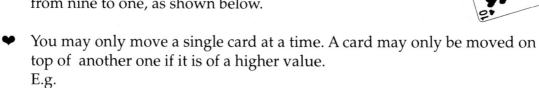

A puzzle for individuals or small groups using sixteen cards (Four Aces, four 2s, four 3s, four 4s).

♥   Try to arrange the sixteen cards into a 4 x 4 array so that the sum of every row, column and diagonal is even.

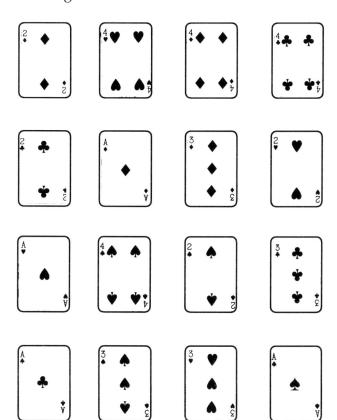

Variation

♥   Play the blind version where the deck of sixteen cards is placed face down and one card is placed on the array at a time. A second card is then turned over and placed in the array and so on. Note once cards are placed they cannot be moved or removed.

Deck Detective

A puzzle for individuals or small groups using three cards.

 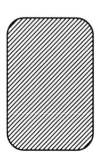 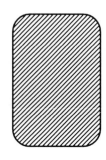

♥   Use the following clues to work out which cards are turned over and the position of the cards.

- ♥   Two of the cards are hearts.
- ♥   Two of the cards are aces.
- ♥   One of the cards is black.
- ♥   One of the cards is a picture card.
- ♥   The two aces are separated by the female picture card.
- ♥   One ace is not a spade.
- ♥   The ace of hearts is on the far left of the set of three cards.

♥   Try this one

- ♥   There are two black cards and one red card.
- ♥   None are picture cards.
- ♥   The three cards are consecutive.
- ♥   The red card is not a heart.
- ♥   Two of the cards are clubs.
- ♥   The smallest card is an eight.
- ♥   The largest card is red.
- ♥   The cards have been placed in descending order from left to right.

♥   Try making some of your own. Start with the answer and work backwards.

Beat That

Materials
You will need a deck of cards with the picture cards removed.

Organisation
A game for two players.

Rules

♥ Deal all the cards in the deck to the two players face down.
(The same as for Snap)

♥ Simultaneously each player turns over their top card.

♥ The player with the bigger number takes the two cards on the table.

♥ If two cards of the same value appear the cards are left on the table to jackpot to the next turn when the winner would pick up four cards.

♥ The winner is the player with the most cards after a set period of time or the player who ends up with all the cards.

Variation

♥ Turn two cards over at a time, add the values to see who has the largest combination. The person with the largest combination picks up all four cards on the table.

Materials
You will need two decks of cards. Separate the picture cards from both decks.

Organisation
A game for small groups or the whole class.

Rules

♥ Each player makes a 4 x 4 array of cards face up.

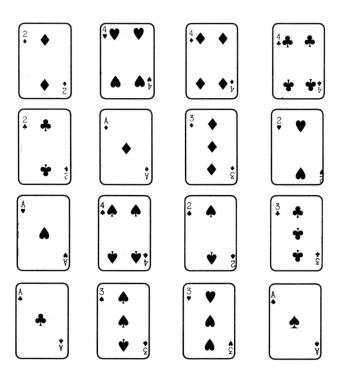

♥ One player takes on the role of caller, flips a card over from the top of the unused deck and calls out the name of the card, eg 7 (It doesn't matter if the card is the 7 of hearts, clubs etc.)

♥ If the card that is called out matches one in the player's array, the player may then turn that card over.

♥ The first player to turn a set of four cards over in a line either horizontally, vertically or diagonally is the winner.

Materials
You will need a deck of cards (Aces = one, Jacks = eleven, Queens = twelve, Kings = thirteen).

Organisation
A game for 2-4 players.

Rules

❧ Each player is dealt four cards face up. The remaining cards are placed in a pack in the centre of the table.

❧ The aim of the game is to be the first player to arrange the cards in either ascending or descending order. This does not have to be in consecutive order, ie 4,5,6,7. It could be 2, 5, 6 and 9. Cards cannot be rearranged – only exchanged.

❧ Starting with the player to the dealer's left each player may exchange one of his/her cards for one from the top of the pack or one from the discard pile. The card which is exchanged is placed into a discard pile.

❧ The first player to arrange his/her cards in order is the winner of that round. The winner receives a point. The first player to accumulate five points is the winner of the game.

  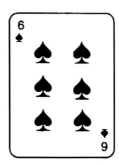 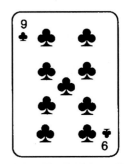

Materials
You will need a deck of playing cards. Picture cards may be removed to allow children to focus purely on numbers.

Organisation
A game for pairs or small groups.

Rules

♥ One player deals five cards to each of the players and leaves the deck face down in the middle of the table.

♥ The player to the left of the dealer begins by asking the next player if he/she has a particular card in his/her hand, e.g. a 4. Note before asking for a card the player must have the matching card in his/her hand.

♥ The player asked must give the card to the player who asks for it if he/she holds it. If the player who is asked does not have the required card then he/she tells the player making the request to "go fish", which means that player must pick up a card from the deck.

♥ If a match is made then the pair of cards must be laid down for all to see. The player does not pick up any more cards and is given another turn. Every time a match is made the player who made the request for a card is given another turn.

♥ The aim of the game is to match all your cards so that none are left in your hand.

Variation

♥ Play Fish +/− 1

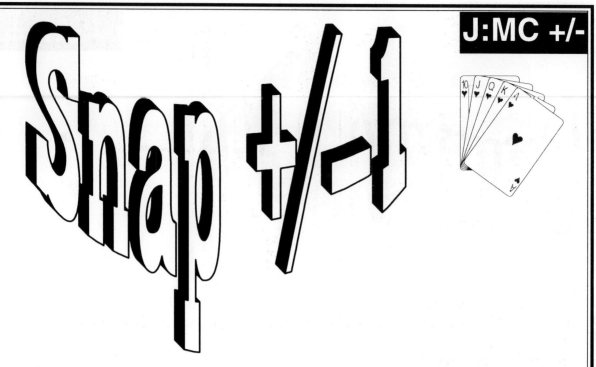

J:MC +/-

Materials

You will need a deck of cards with the picture cards removed. Ace may equal one or eleven.

Organisation

A game for two players.

Rules

♥   The game is played along similar lines to 'snap'.

♥   One player deals all the cards face down to the players.

♥   Each player turns over their top card. Instead of slapping the pile of cards when the values on the two cards match, the pile of cards should be slapped when the values differ by one. For example if a 7 is placed on the pile and then an 8 is discarded on top a player may slap the pile and pick up all the cards. If an 8 was on the pile and a 7 was discarded then the pile of cards could also be slapped.

♥   The winner is the player with the most cards after a set period of time or the player who ends up with all the cards.

Variations

♥   Play the standard game of snap to develop number recognition.

♥   Play Snap +/− 2. i.e. snap when the values differ by two.

Materials
You will need a deck of cards with the picture cards removed.

Organisation
A game for pairs or small groups.

Rules

♦ Prior to starting the game a target number should be chosen (e.g. 15)

♦ Each player is dealt 5 cards.

♦ Four cards are dealt face up and the remaining deck placed in the middle.

♦ Players take turns to place one of their cards on one of the four cards that is face up and add the values to try to reach the target number. Depending on the size of the target number, players may place more than one card on a single pile. Cards may only be laid down if the exact total can be produced. A player's turn is over after he/she produces the target number or chooses a card from the deck.

♦ Players choose a card from the deck if they cannot lay down a card or cards.

♦ Players reaching the target get to keep the cards in a separate pile. The values of these cards are added at the end of the game to determine a winner.

♦ Once a pile is removed a card is turned over from the deck to replace it.

Variations

♦ Change the target number.

♦ Set a target to be reached by multiplication rather than addition.

Make 10

Materials
You will need a deck of playing cards with the picture cards removed.

Organisation
A game for pairs or small groups.

Rules

♥ One player deals out ten cards in a row.

♥ The first player then looks across the row of cards for a combination of cards (any number of cards is fine) that adds to make ten e.g. 6 + 4, 7 + A + 2.

♥ Only one combination may be removed. The aim of the game is to collect as many cards as possible, so combinations that require more cards are favoured.

♥ Once a combination of cards has been removed the cards are replaced by the dealer with new ones from the pack.

♥ Play continues until there are no more cards or until players can no longer make up combinations that add to ten. Players then count their cards to determine the winner.

Variation

♥ Choose a different target number eg twelve.

Materials

You will need a deck of playing cards with the 10s and picture cards removed.
Ace = one.

Organisation

A game for one or two players.

Rules

♥ One player deals all 36 cards, face up in a 3 x 3 array.

♥ There should be four cards in each pile.

♥ Players take turns to pick up any number of cards, which when added make 10. As cards are taken from the pile a new card is revealed underneath.

♥ Play continues until all the cards have been used or until no more combinations that add to ten can be made.

♥ The winner is the player with most cards at the end of the game.

Variation

♥ Choose a different target number eg twelve.

Materials
You will need a deck of cards with all the picture cards removed for each player.

Organisation
A game for the whole class.

Rules

♣ Each student shuffles his/her deck and lays it face down on the desk.

♣ The teacher calls out "go" and then the students flip over one card at a time, keeping a running total.

♣ After thirty seconds, one minute or two minutes, depending on the ability of the class, the teacher says "stop".

♣ The players then record the total they reached and the number of cards flipped in order to reach the total.

DATE	TIME	CARDS	TOTAL	CHECKED
4/7	30 sec	8	37	✔

♣ Flipped cards are then handed to the closest player who checks they add to the stated total.

♣ Children can build up a chart similar to the one above to monitor their own performance.

Variation

♣ Remove the 7, 8, 9 and 10s for younger players.

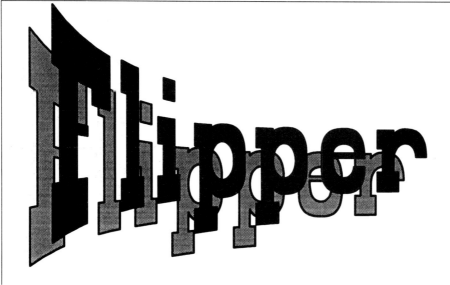

DATE	TIME	CARDS	TOTAL	CHECKED

I Spy

Materials

You will need a deck of cards with all the picture cards removed.

Organisation

A game for two players.

Rules

♣   The forty cards are dealt face up in a 10 x 4 or 8 x 5 array.

♣   One player challenges the other player to find two cards next to each other that add to make a particular number by saying "I spy with my little eye two cards which add to make _____"

♣   The other player then looks for two cards that are next to each other either horizontally or vertically that add to make the number and then picks this pair of cards up and any other pair next to each other that add to make the stated number.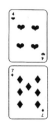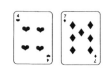

♣   If the second player misses any pairs that add to the chosen number, then the first player may claim them.

♣   Players swap roles and continue until the table is cleared.

♣   The winner is the player with the most cards at the end of the game.

♣   As large gaps appear the size of the array may be reduced to help fill the gaps.

Variations

♣   Allow children to add three cards together.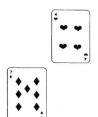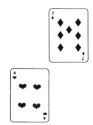

♣   Allow children to use pairs of cards that are diagonally opposite each other.

♣   Change operations: i.e. Use subtraction or multiplication.

Materials
You will need a deck of cards (the ace counts for eleven, picture cards are worth ten).

Organisation
A game for 2-4 players.

Rules

The aim of the game is to make a total of thirty one or a total greater than the other players.

♣ Each player is dealt three cards. One card is placed face up in the centre of the table (this forms the discard pile) and the remainder of the deck is placed next to it.

♣ The player to the left of the dealer starts by either drawing a card from the deck or from the discard pile and then discarding one from his/her hand.

♣ Play continues in this fashion until a player can make thirty one, exactly, by adding the values of the cards in his/her hand

OR

until one player knocks the table. By knocking the table the player indicates he/she is happy with his/her total. The other players have one more turn and then all hands are exposed, totalled and compared.

♣ The winner of the round is the player with the highest total.

♣ Players keep a cumulative record of their totals. The first player to reach one hundred wins the entire game.

Variations
♣ Discard before picking up.
♣ Remove cards with lower values.

Materials
You will need a deck of cards with the 10s and picture cards removed.

Organisation
A game for pairs or the whole class.

Rules

♣   The deck is shuffled and left face down on the table.

♣   Players take turns to pick a card from the top of the deck and turn it over.

♣   The player must then place his/her card in front of them in either the ones, tens, hundreds or thousands place. The card must be placed before another is drawn from the deck.

♣   The winner is the player who produces the largest number.
For example 6 741 was produced using a 6, 7, 4 and ace. 7 641 was the best number that might have been formed.

Variations

♣   Make larger/smaller numbers. Include decimals.

♣   Use a scoring system where the person with the lowest amount scores zero while all the other players score the difference between their number and the lowest number.

Materials
You will need a deck of cards with the picture cards removed.

Organisation
A game for 2 – 4 players.

Rules

- All the cards are dealt.

- Players must not look at their cards.

- The dealer chooses a number from one to ten e.g. three. This becomes the multiple for that round.

- The players then take turns to flip over their top card and place it on a pile in the middle.

- Players must keep a mental running total of the cards (i.e. the values of the cards) added to the pile.

- When the total is a multiple of the chosen number – say three – then the first player to call out 'three three three' collects the cards.

- Other players may challenge whether the result is really divisible by a that number. A calculator may be used to settle disputes. If the player calling out "three three three" was wrong then he/she must place two ofhis/her cards in the centre and play continues until the multiple comes up.

- The winner is the person with the most cards in his/her hand after a set time or the only player with cards left in his/her hand.

Variations
- Use different multiples, e.g. 5, 6.

- Use pairs of multiples, e.g. 2 and 3 or 2 or 3. Note it is easier to find multiples of 2 *or* 3 rather than both together.

Materials
You will need a deck of playing cards with the picture cards removed. The deck then needs to be split into two packs; one containing three sets of cards ace to 10 and one which contains one set of cards ace to 10. Ace = one.

Organisation
A game for one to three players.

Rules

♥ The player(s) place the two packs (one containing thirty cards and one containing ten cards) side by side.

♥ One card from the ten card pack is flipped over. The value shown on the card indicates the table to be practised for the round.

♥ Cards are then flipped over from the pack of thirty and the product of the two cards is calculated.

♥ Players keep the cards for the calculators they correctly answer.

♥ Play continues until the pack of thirty is exhausted.

♥ The pack of thirty is then shuffled and placed face down on the table.

♥ Another card is flipped over from the pack of ten to set the table to be practised for the second round.

Variation

♥ Add the values on the card instead of multiplying them.

♥ Give each player a set of ten cards instead of using one pile of thirty.

Getting Closer

Materials
You will need a deck of cards with the picture cards removed.

Organisation
A game for 2 – 4 players.

Rules

❧ Deal four cards to each player.

❧ Turn up two cards from the deck. The first represents the tens and the second, the units. This becomes the target number.

❧ The players now turn over their cards and try to form two, two-digit numbers that when added or subtracted will be as close to the target number as possible.

  +

❧ Players score by finding the difference between their total and the target number.

❧ Play continues for several rounds. The winner is the player with the smallest total.

Variation

❧ Players should try to produce a total as far away from the target as possible.

99 OR BUST

Materials
You will require a deck of cards.

Organisation
A game for 2 – 4 players.

Rules

♠ Aces = one, Picture cards = minus ten (i.e. subtract 10 from the total), other cards are counted according to their face value.

♠ Each player is dealt three cards and the balance of the pack is placed face down on the table.

♠ Play commences with the player to the left of the dealer and continues in a clockwise direction.

♠ In turn each player draws a card from the pack and then discards one from his/her hand.

♠ A running total of the cards on the table is kept.

♠ The aim of the game is to reach ninety-nine or force an opposing player to discard a card which makes the total higher than ninety-nine.

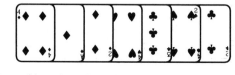

e.g. (4 + 1 + 2 + 4 + 3 + 4 + 2 = 20)

♠ As the total gets closer to ninety-nine the game becomes very exciting and various tactics and strategies come into play.

Materials
You will need a deck of cards. Ace = one and picture cards equal ten.

Organisation
A game for 2 – 4 players.

Rules

♦ One player draws three cards from the deck and lays them face up on the table.

♦ The first player to call the product of the three cards is awarded that number of points.

♦ No points are awarded for an incorrect answer.

♦ If three cards of the same suit or three cards of the same value turn up then the first player to call out "HIT THE DECK" and correctly multiply the three numbers receives **double** points.

Variations:

♦ Draw two, four or five cards.

♦ Use addition instead of multiplication.

On Target

Materials

You will need a deck of cards with the picture cards removed. Ace = one.

Organisation

A game for small groups.

Rules

♠ The dealer chooses a two digit target number and then deals five cards to each player. Play begins with the first player to the left of the dealer and continues in a clockwise direction. In turn each player lays one card from his/her hand face up on the table.

♠ A running total of the cards is kept until no player can lay down a card without exceeding the target number.

♠ Once the game has concluded the players add the values of the remaining cards in their hands to find their scores.

The winner is the player with the least score after three rounds.

Sample game

A target of **52** was chosen and the four players played the following cards:

Player 1	Player 2	Player 3	Player 4
9	9	10	8
6	5	Ace	2
Ace	Can't go	Ace	

52

Cards Remaining in Hand			Score
Player 1	4	3	7
Player 2	4	3	2	9
Player 3	9	7		16
Player 4	8	6	5	19

> To work out the score for that round simply total the value of the cards left in the player's hand.

Therefore Player 1 is the winner for that round. The winner is of the game is the player with the least score after three rounds.

Variations

♠ Use different target numbers.
♠ Remove larger cards and reduce target number for younger children.
♠ Remove the odd or even numbered cards from the pack.

Materials

You will need a deck of playing cards with the picture cards removed. Aces = one.

Organisation

A game for two players.

Rules

♠ Deal out half the cards to each player.

♠ Both players lay out a card face up. The first to call the product picks up the two cards.

♠ Ties are settled by leaving the cards on the table. The winner of the next call picks up all the cards on the table.

♠ The winner is the player with the most cards once all the cards have been used.

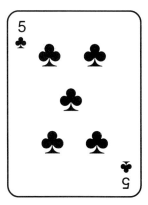  

Variations

♠ Use addition or subtraction.

♠ Remove cards which are beyond the children's ability e.g. 7, 8, 9.

Materials
You will need a deck of cards with all the picture cards removed.

Organisation
A game for a small group.

Rules

♠ One player deals out two cards face down and one card face up to each player. (The face up card is the addition card.)

♠ The dealer then states either **High** or **Low** and turns over his/her cards. These cards are multiplied and the number on the third card is added to the product.

e.g.

$(5 \times 7) + 3 =$

♠ The other players now turn over their cards and work out their totals. If a player scores less than the dealer, when the call is **Low** then he/she earns a point. If the call was **High** and the player scored less than the dealer then he/she does not score.

♠ The winner is the player with the highest score after ten rounds.

Materials

You will need a deck of playing cards. The picture cards all have a value of ten. Ace = one or eleven.

Organisation

A game for pairs or small groups.

Rules

♣ The dealer chooses a two-digit target number and deals three cards to each player. The player to the left of the dealer tries to make the target number using his/her three cards and any of the four mathematical operations (addition, subtraction, multiplication or division). If the player cannot make the number one card is discarded from the hand and another one drawn. Play continues in a clockwise direction.

♣ The winner is the player who is able to make the target number with his/her three cards.

Sample game: *Target number 32*

Player 1	J	3	4	$10 \times 3 + 4$	= 34
Player 2	9	8	4	$9 \times 4 - 8$	= 28
Player 3	Q	8	4	$10 \times 4 - 8$	= 32 (Win)

Variations

♣ The size of the target number and/or the operations used can be altered depending on the age of the players.

♣ Deal out more than three cards.

This game provides an ideal context in which to discuss "rule of order".

Materials

You will need a pack of playing cards with the picture cards removed.

Organisation

A game for two players.

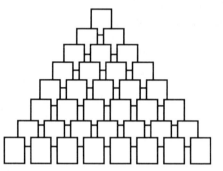

Rules

❧ The players arrange thirty six cards face up to form a pyramid (as shown) with eight horizontal rows each one overlapping the one above it.

❧ The rest of the pack (four cards) is placed on the table face down. This is called the leftovers pile.

❧ A player begins by removing from the bottom row up to five cards which when totalled form a multiple of ten. These cards are set aside for scoring at the end of the game. Player number two may then remove any uncovered cards to form a multiple of ten.

❧ A card may not be removed if it is covered by another card.

❧ Should a player be unable to find uncovered cards to total a multiple of ten, the top card in the leftovers pile is turned over. The card from the leftovers pile may be used in combination with any uncovered cards to form a multiple of ten.

❧ If this card cannot be used in this way the next card is also turned; this and other uncovered cards may now be combined to form a multiple of ten.

❧ This process is repeated, if necessary, until all four cards in the leftover pile have been used.

❧ The game ends when all the cards in the pyramid have been used, or when no combinations of up to five cards will give a multiple of ten.

❧ The winner is the player with the larger total when the cards are added up.

Materials

You will need to separate the four aces (Aces= one), four 2s, four 3s and four 4s from a deck of cards.

Organisation

A game for pairs

Rules

♦ The dealer lays the cards out in the following manner.

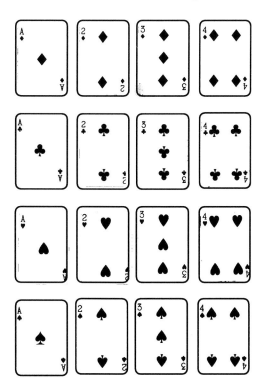

♦ In turn each player selects a card from the sixteen laid out cards and turns it over. Once it has been turned over it may not be reused. The value is noted and a combined running total is kept. The player who makes the total equal to twenty one is the winner.

♦ Any player going over twenty one busts and loses the game.

Materials
Each player will need a set of ten playing cards of any suit from ace to 10. Ace = one.

Organisation
A game for pairs.

Rules

♣ Each player is dealt ten cards.

♣ Each player arranges their ten cards in five pairs but keeps these pairs secret from the other player.

♣ He or she writes the product of each pair on the recording sheet.
Pair 1

$$\boxed{9} \ \times \ \boxed{7} \ = \ \bigcirc{63}$$

♣ The second player must then try to guess which pairs produce the given product.

♣ Players then swap roles.

♣ The winner is the player who correctly guesses the most pairs.

Secret Pairs

Pair 1   [ ]   X   [ ]   =   ( )

Pair 2   [ ]   X   [ ]   =   ( )

Pair 3   [ ]   X   [ ]   =   ( )

Pair 4   [ ]   X   [ ]   =   ( )

Pair 5   [ ]   X   [ ]   =   ( )

Materials

You will need a calculator and a deck of cards with the tens and picture cards removed.

Organisation

A game for one or two players or whole class.

Rules

❧ Deal the cards face down into four piles. Each pile represents a particular value i.e. thousands, hundreds, tens or units.

❧ Flip the top card over from each pile and enter the number onto the calculator e.g. 7219.

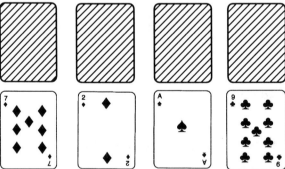

❧ One player flips another card over and the second player must change the number on the display of the calculator using addition or subtraction.

| The player would need to add 300 to 7219 to turn it into 7519. | | | | |

❧ Play continues for five card flips and then players swap roles. Alternatively play can continue until the player with the calculator makes a mistake.

Variations

❧ Use five or six piles of cards to represent larger numbers. Use decimals.

❧ Flip more than one card at a time.

CARD CALCULATIONS

Materials
You will need a deck of cards with the 10s and picture cards removed.

Organisation
A game for a small group.

Rules

❧ Each player is dealt four cards face up.

 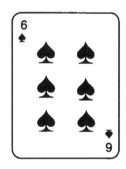  

❧ Each player then tries to make a number sentence which gives a single digit answer.

❧ The answer becomes the score for that player.

e.g.
$7 + 3 + 2 - 6 = 6$ 6 points
$6 + 7 - 3 - 2 = 8$ 8 points
$36 - 27 = 9$ 9 points

The best combination of numbers is the one that produces the largest number and hence the largest score.

❧ The winner is the player with the largest score after five rounds.

Variations

❧ Aim to produce the lowest score.

❧ Deal out more/less cards.

Answers

Symmetrical Cards. p. 5
All of them. 3, 5, 9. 3, 5, 9

Spotted Cards. p. 6
If we consider a single suit.
Each picture card has 4 spots (3 x 4 = **12**)
Every number card has one spot directly under each number (2 x 10 = **20**)
Each card has the same number of spots as its face value.
(1 + 2 + 3 +4 + 5 + 6 + 7 + 8 + 9 + 10 = **55**)
Therefore there are 87 spots in one suit. There are 4 suits, therefore there are
4 x 87 or 348 spots on an entire deck.

Cut and Predict. p. 7
About half would be red. 25, 50, 100. 10, 15, 25.

Magic Cards. p. 8.
They always total to 45.

Boxed Cards. p. 9
One solution:

6	10	2
7		3
1		8
4	9	5

Classifying Cards. p. 10, 11
There are 26 red cards, 2 red aces and 4 aces
There are ten odd numbered red cards, ten even numbered red cards and six red
picture cards. There are ten odd numbered black cards, ten even numbered
black cards and six black picture cards.
1. Ace of Hearts.   2. J, Q, K of Diamonds   3. Ace, 3, 5, 7 & 9 of Clubs.
4. J, Q, K of Spades   5. J, Q, K of Clubs   6. 10 of Diamonds
7. 4 of Spades, 4 of Clubs.

Card Conundrum. p. 12

1)	K	Q	J	A	2)	K	J	A	Q	3)	A(h)	K(c)	Q(d)	J(s)
	Q	K	A	J		A	Q	K	J		Q(s)	J(d)	A(c)	K(h)
	J	A	K	Q		Q	A	J	K		J(c)	Q(h)	K(s)	A(d)
	A	J	Q	K		J	K	Q	A		K(d)	A(s)	J(h)	Q(c)

Card Puzzle. p. 15
There are several solutions. One is shown on page 15.

Deck Detective. p. 16
1.   Ace of Hearts, Queen of Hearts, Ace of Spades.
2.   10 of Diamonds, 9 of Clubs, 8 of Clubs.